Oxford CONNECTIONS

Interdependence and Adaptation

Julia Bruce

Series editor **Sue Palmer**

OXFORD
UNIVERSITY PRESS

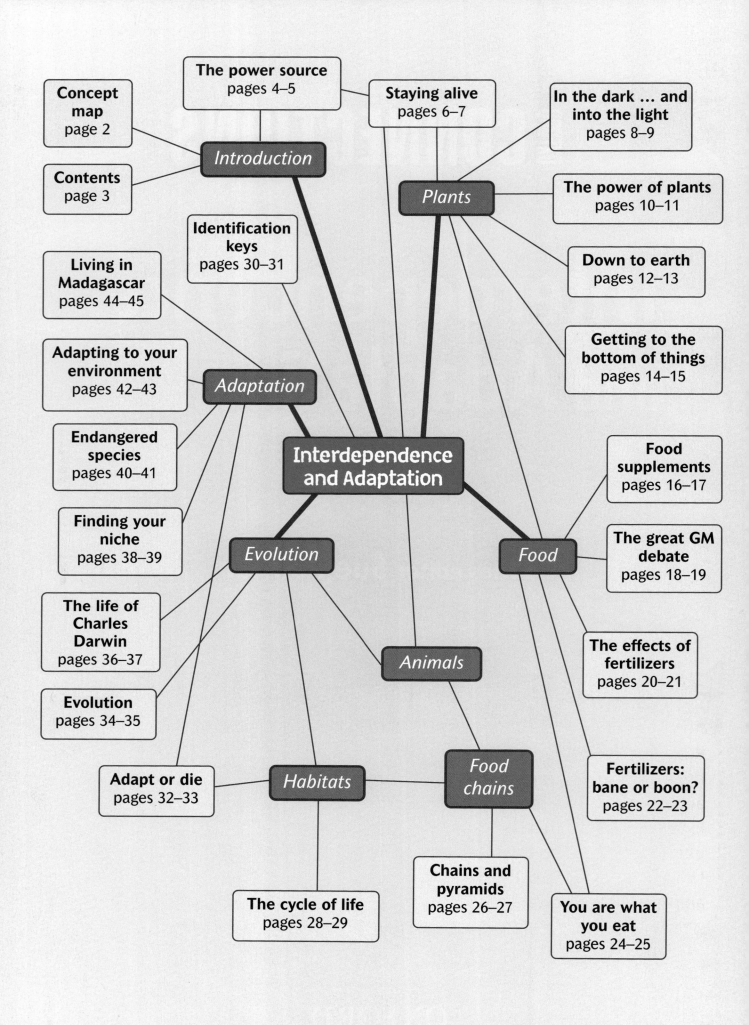

Introduction

Plants

Interdependence and Adaptation

Adaptation

Evolution

Animals

Food

Habitats

Food chains

Contents

Read this book and discover how all living things, from the smallest plant to the largest animal, depend on each other for their survival. Find out how different types of living things are uniquely adapted to their own environment. I was particularly fascinated by the peculiar animals of Madagascar.

Dr Mike Kent
Head of
Zoological Programmes,
St Austell College

The power source

What is energy?

We use the term energy to mean all sorts of things. If someone is lively we say they have lots of energy. We also use energy to mean movement or vigour. In science, to have energy means the ability to make something move. Energy drives all the processes that go on both on the Earth and in space.

Why do we need it?

Without energy there could be no life. Animals and plants need energy to keep their life processes, such as growth, movement and reproduction, going. Most organisms on Earth get that energy ultimately from the Sun.

The Sun is the ultimate source of energy for almost all life on Earth.

Biofuels

Burning wood or charcoal releases energy that the trees once took in from the Sun. Methane, a gas produced when organic matter decays, and biodiesel, from palms and other plants, can also be used as fuel.

Plants

Plants use the Sun's energy to turn water and carbon dioxide into new growth.

Fossil fuels

Oil, coal and natural gas are all made from the fossilized remains of ancient plants and **microorganisms**. When we burn them, we release the energy stored in the original organisms.

Food

Energy from the Sun is harnessed by plants and passed on to the animals that eat them. We humans get energy from both animal and plant food sources.

What is the Sun?

In ancient times people thought that the Sun was a god, often shown in pictures or carvings riding a golden chariot across the sky. Today, we know that the Sun is, in fact, our nearest star. It is 1,392,000 km wide and about 4.6 billion years old. It shines and provides heat because it is constantly burning up the massive stores of hydrogen gas in its core. In about 5 billion years, it will cool and expand, becoming an enormous star called a red giant, before it finally shrinks into a small star called a white dwarf. Then it will gradually fade away and die as it uses up the very last of its fuel.

The Egyptian Sun god Ra-Harakhte

Animals

Animals cannot obtain energy directly from the Sun. Instead, by eating plants they acquire some of the solar energy the plants have harnessed.

Most of the energy that reaches the Earth from the Sun is lost by being reflected back into space.

Staying alive

All living things need food to support life, provide energy and promote growth. Plants and animals obtain food in very different ways.

Plants

- use the Sun's energy directly to make own food

- are called 'producers' because they produce food

- produce food in special cell structures called chloroplasts – found mainly in the leaves

- use the Sun's energy to combine water and carbon dioxide (a gas from the air) into a sugary food – carbohydrate
- process is called **photosynthesis**

  sunlight

  carbon dioxide + water ⟶ carbohydrate + oxygen

- obtain other **nutrients** from soil

- store food in the form of **starch** (much as animals store fat in their bodies)

- have sap which transports food and nutrients to where they are needed for growth and development

light energy

oxygen

food

carbon dioxide

water and minerals

Plants use sunlight to make food.

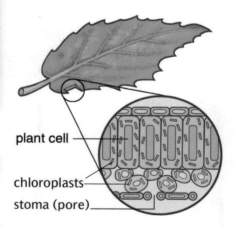

plant cell

chloroplasts

stoma (pore)

Cross-section of a leaf

Did you know that a typical plant cell contains 50 chloroplasts?

Grazing animals, such as zebras, obtain energy from plants.

Animals

- cannot make own food within their bodies

- are called 'consumers' because they consume food energy made by plants – the 'producers'

- obtain energy from eating plants or other animals

- digestion breaks down food and respiration releases the Sun's energy stored in it

Are humans producers or consumers?

In the dark ...

What happens if plants are kept in the dark?

You will need:
- 2 pot plants
- watering can

Do plants need light to thrive? Try this experiment.

What has happened here? What are the differences between the two plants? Does Plant B look healthy?

What will happen to the plant that was kept in the dark if it is returned to the light? Leave your plant on the window sill for a few days and see what happens.

Plants need light to grow healthily.

Why are plants green?

Plants are green because they contain a green pigment called **chlorophyll**, which is found in chloroplasts. Chlorophyll absorbs light energy from the Sun for **photosynthesis** (food-making process.)

Even plants that don't look green still contain chlorophyll.

Copper beech leaves

... and into the light

How can we show that plants use sunlight to make their own food?

You will need:

- a strong, leafy plant e.g. a geranium
- black plastic sheeting e.g. dustbin liners
- tape
- scissors
- saucepan
- saucer
- heat-proof beaker (500 ml)
- iodine solution
- methylated spirit
- tweezers
- thread (2 different colours)

What you need to know

Iodine solution is a yellow–orange liquid that stains things purple–black if they contain **starch**.

1 Cover some of the leaves of the plant with a thick layer of black plastic, securely taped so no light can get in.

2 Water the plant and leave it on a window sill for two days.

3 Choose two leaves, one that has not been covered up and one that has. Tie a coloured thread on each stalk to show which leaf is which.

4 Put 100 ml of methylated spirit into the beaker and place in a saucepan of water (about one third full). Heat gently until the methylated spirit begins to boil.

5 Take the pan off the heat. Drop the leaves into the hot water for about a minute.

6 With the tweezers, remove the leaves and put them into the beaker of methylated spirit. Leave them until they turn white.

7 Carefully take the leaves out of the beaker with the tweezers and put them into a saucer. Pour a little iodine solution on to each leaf.

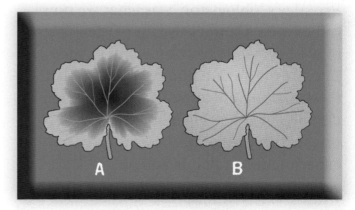

A   B

Results and Conclusion		
	Leaf A (kept in light)	**Leaf B** (kept in dark)
Observation	Leaf stained purple–black	Leaf stained yellow
Conclusion	Starch present. Photosynthesis has taken place.	No starch present. No photosynthesis has taken place due to absence of sunlight.

The power of plants

The discovery of **photosynthesis** can be traced back to the 18th century, a time when scientific understanding was very different from what it is today. In the 1770s people were only just beginning to realize that there was something special in the air that kept animals alive. The English chemist, Joseph Priestley, called it 'phlogiston'. He noticed that air in a small space eventually became unbreathable if a candle was burned in it. Priestley thought this was because the phlogiston had been used up or removed. However, he found that, if a plant was put into the space, the same air eventually became breathable again.

Joseph Priestley

Priestley's 'phlogiston' was what we now call oxygen. Candles use up oxygen when they burn, while plants give off oxygen into the air during photosynthesis (their food-making process). In 1779, a Dutch doctor called Jan Ingenhousz investigated this further and published his own experiments on plants and air.

> It will, perhaps, appear probable, that one of the great laboratories of nature for cleansing and purifying the air of our atmosphere is placed in the substance of the leaves, and put in action by the influence of the light …
>
> Jan Ingenhousz 1779

The life of Jan Ingenhousz

JAN INGENHOUSZ was born in Breda, Holland in 1730. He was the son of a leather merchant. Jan was a clever boy and he ended up going to university to study medicine, first in Holland, then in Paris and Edinburgh.

He worked as a doctor in his home town until he was 35, when he went to work in a hospital in London. Here he was one of the pioneers of inoculation* against the deadly disease, smallpox. In 1768 his work was noticed by the king, George III, who sent him to Vienna to inoculate the Austrian Empress Maria Theresa. Ingenhousz stayed in Austria as court physician until 1779, when he returned to England.

While he was in Austria he studied other subjects, including electricity and plants. His work on plants was very detailed and he did hundreds of experiments to try and find out exactly how they worked.

Ingenhousz had met Priestley when he was working in London and knew all about Priestley's work on air. Ingenhousz wanted to find out more about the effect plants had on air, so he set up experiments to investigate the gases plants give off under water.

* Inoculation is a kind of vaccination

He knew that submerged plants produce small gas bubbles but that if the plants were put in the shade, the bubbles eventually stopped. Ingenhousz collected the gas given off by a submerged plant. He held a glowing taper in the gas, and it began to burn again. This suggested that the air was full of what we call oxygen. In the dark, he found that the plants gave out less gas. When he collected this gas, he found that it could put out a flame, meaning that the air was rich in carbon dioxide. His experiments proved that plants produce oxygen in the light, but in the dark they give out carbon dioxide.

Ingenhousz performed about 500 experiments on plants and published his findings in 1779 in a book called, *Experiments Upon Vegetables, Discovering Their Great Power of Purifying the Common Air in Sunshine, and of Injuring It in the Shade and at Night.*

Jan Ingenhousz

In his book he showed that:

- plants, like animals, breathe all the time
- in the light, plants make oxygen (he called this 'the restoration of air')
- only the green parts of the plant do this
- plants give off carbon dioxide when they are in the dark. This 'damages' the air, but plants restore the air far more than they damage it.

In effect, Ingenhousz had discovered photosynthesis, although he did not name the process, nor did he understand exactly how it worked. He also realized that animals and plants are interdependent: animals (including humans) need oxygen from plants, and plants need the carbon dioxide people and other animals breathe out.

As well as his work on plants and inoculation, Ingenhousz investigated heat conduction and electricity, inventing a machine to create static electricity. He returned to England in 1779 to publish his book and died in 1799 at Bowood in Wiltshire.

Down to earth

Take a look at some soil under a microscope, or a magnifying glass, and you will find that it is not just dirt or mud. It is made up of all sorts of different things, including rock fragments (stones, gravel, sand, silt or clay), decomposed plants, bits of dead animals such as beetles, worms and spiders, and even **bacteria** and fungi.

There are many different types of soil, depending partly on the type of rock that the soil has formed on. Some plants thrive only on a particular sort of soil. If you live on chalky soil, for instance, you will find that azaleas won't grow well in your garden. They prefer a more acidic soil with a higher **humus** content.

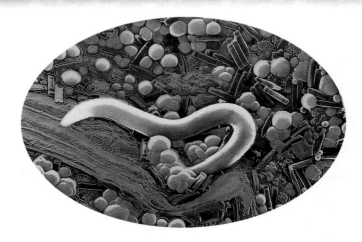

A soil sample viewed under a microscope.

A good fertile soil will contain:
45% mineral particles (clay/sand/gravel);
5% organic matter from decomposing plant and animal material (humus);
25% water; and 25% air spaces.

Plants for different soils

Soil type	Rock type	Drainage	Comments	Typical plants	
Sandy	Sandstone, granite	Excellent	Dry, low humus content, acidic, less than 20% clay	Conifers, heathers	heathers
Chalky	Chalk	Good	Dry, low **nutrients**. Flint pebbles often found in chalk soil	Cotoneaster, hazel, orchids	hazel
Clay	Mudstone, clay	Poor	Wet, sticky soils that clump together. At least 30% clay particles	Ferns, bog myrtle, guelder rose	ferns
Humus-rich	Variable	Good	Often old soils, high humus content. Often acidic	Hydrangeas, rhododendrons, azaleas, camellias	hydrangea

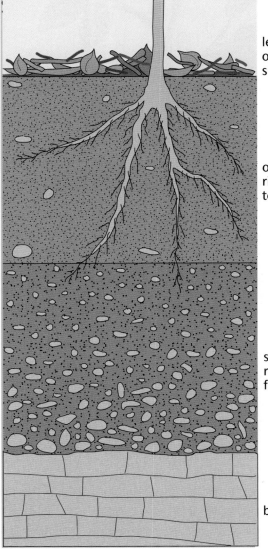

leaf litter on surface

organic-rich topsoil

subsoil of rock fragments

bedrock

Cross-section of soil showing structure

Do plants need soil?

Because we see plants growing naturally in soil, it seems obvious that plants need soil to survive. It is certainly true that most plants get water and many of their nutrients from soil, but plants can actually survive quite happily without it, as long their roots are supported by some sort of solid material, such as gravel, and provided with nutrients in sufficient quantities. The method of growing plants, without soil, in gravel and water enriched with nutrients, is called 'hydroponics'.

A hydroponic system for growing plants

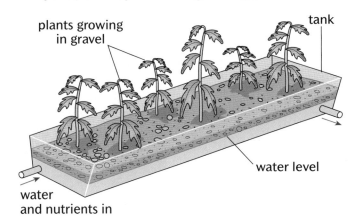

plants growing in gravel

tank

water level

water and nutrients in

Living in the soil

Healthy soil teems with organisms that are invisible to the naked eye. These **microorganisms** include bacteria, fungi, nematodes (tiny worms), protozoa (animals consisting of only one cell) and viruses. Microorganisms are vital to many soil processes.

Microorganisms:

- break down, or decompose, organic matter in soil releasing nutrients for plants
- convert nitrogen into a form useful to plants
- bind the soil together and improve its texture
- give fresh, moist soil its distinctive smell.

Getting to...

Roots have several important functions. As well as anchoring the plant in the soil and absorbing water and **nutrients**, they often also act as food stores. Think of root vegetables: carrots, parsnips, turnips are all 'tuberous' roots that store **starch** for their parent plant. Some roots have other specialist functions. Buttress and stilt roots, for instance, provide extra support in shallow or boggy soil. Mangrove plants, which live in shallow coastal waters in some tropical countries, have roots that poke out above the salty water and help the plant take in oxygen from the air.

There are two main types of root systems

1 Tap roots

e.g. carrots, radishes, parsnips and dandelions. They have a main root from which other side roots grow.

2 Fibrous roots

e.g. grasses, marigolds, and beans. They have a spread of roots of similar size with no obvious main root.

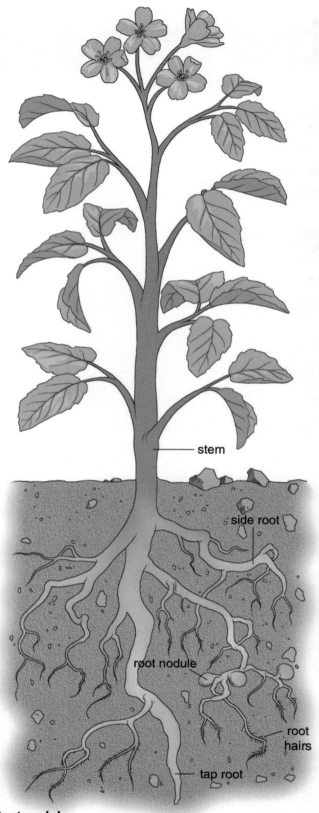

stem

side root

root nodule

root hairs

tap root

Root nodules
Lumpy growths in certain plants e.g. bean, alder, containing **bacteria** that take nitrogen from the air and convert it into fertilizer.

Root hairs
Small, hair-like extensions of the outer root cells. Water and nutrients are absorbed through root hairs.

Tap root
Enlarged central root that grows downwards (other roots grow from it).

the bottom of things

Specialist aerial roots

Some plants have special roots which grow above ground.
As well as taking in water and nutrients, they can help to
hold up the tree, or obtain oxygen.

Root type	Description	
Breathing roots	• found on plants in swampy areas where there is not enough oxygen in the water for the root to survive • grow above the water or mud and are almost hollow, allowing air to pass directly into the root tissue	
Buttress roots	• board-like growths on base of large trees in shallow soils • help support tree, like buttresses that support castle walls	
Clasping roots	• found on plants such as poison ivy to attach the plant to other trees – also absorb nutrients from surface of tree • sometimes grow to the ground	
Prop roots	• grow above ground from base of stem or branch and act as support	
Stilt roots	• support trees in swampy areas such as mangroves during changes in mud level (Mangroves are marshy areas along the coast in tropical climes.)	

Food supplements

A BALANCED DIET FOR YOUR PLANTS

Is your breakfast cereal 'fortified with vitamins and minerals'? Do you sometimes take food supplements? Do you know anyone expecting a baby? Perhaps she takes supplements to help the development of her growing foetus. We supplement our diet with vitamins and minerals to keep us healthy without giving it too much thought. But it's not only humans that need extra **nutrients** – plants do too. To ensure your house and garden plants are healthy you need to make sure they have a balanced diet. If they have discoloured patches on their leaves, are not growing, flowering or fruiting well, the chances are they are short of one or more essential nutrients . . .

Cereal Nutrition		
Thiamin (B1)	0.4mg (29% RDA)	1.2mg (86% RDA)
Riboflavin (B2)	0.6mg (41% RDA)	1.4mg (88% RDA)
Niacin	4.7mg (26% RDA)	15.3mg (85%RDA)
Vitamin B6	0.6mg (30%RDA)	1.7mg (85%RDA)
Folic Acid	59μg (29% RDA)	170μg (85% RDA)
Vitamin B12	0.8μg (77% RDA)	0.9μg (90%RDA)
Pantothenic Acid (B5)	1.9mg (32%RDA)	5.1mg (85%RDA)
Iron	3.6mg (26%RDA)	11.9mg (85%RDA)

Plants get most of the extra nutrients they require from the soil. They need:

✿ *iron* and *copper* to make **chlorophyll**

✿ *nitrogen* to build their cell walls and make proteins

✿ *potassium* to help control their intake of water and build up their resistance to disease

✿ *phosphorus* to help **photosynthesis** and reproduction

However, sometimes the soil is short of one or more of these vital nutrients. In these cases, the plants will not be able to grow well, and may eventually die.

If plants don't get the very small amount of *copper* they need, their flowers may not develop properly and they may get infections. The leaves of plants which have not had enough copper may discolour or grow twisted. Growth and fruit production is stunted.

Plants that lack *iron* turn yellow (chlorosis) often starting at the tips of shoots.

How to have healthy plants

To make up for any shortage of nutrients, plants can be given the equivalent of a food supplement in the form of fertilizers and feeds. Fertilizers contain all the important nutrients that plants need. Gardeners and farmers add them to the soil, so that plants can take them in through their roots and stay healthy.

ANALYSIS
NPK FERTILISER 15:3015 WITH TRACE ELEMENTS

Total Nitrogen (N)	15%	Iron (Fe) chelated by EDTA	
Phosphorous Pentoxide (P₂O₅)		soluble in water	0.15%
soluble in neutral ammonium		0.15% chelated by EDTA	
		Manganese (Mn)	
citrate and in water	30%(13.1%P)	chelated by EDTA	
of which soluble in water	30%(13.1%P)	soluble in water	0.05%
Potassium Oxide (K2O)		0.05% chelated by EDTA	
soluble in water	15% (12.4%K)	Molybdenum (Mo)	
Boron (B) soluble in water	0.02%	soluble in water	0.0005%
Copper (Cu) soluble in water	0.07%	Zinc (Zn) soluble in water	0.06%

Signs of nutrient deficiencies

Plants show distinctive symptoms when they are short of nutrients. This chart shows the signs to look out for.

Symptom	Likely nutrient deficiency
Yellowing	iron, nitrogen
Other leaf discoloration	copper, phosphorus
Leaves dying	copper, potassium
Stunted growth	nitrogen, potassium, phosphorus
Lack of new growth	nitrogen
Deformed leaves	copper
Poor fruiting	copper, phosphorus
Vulnerability to disease	potassium, copper

plant does not get ough *nitrogen*, its ms are thin and indly, and its growth stunted. Most of the itrogen available goes o the new leaves, so the ld leaves do not get enough and they turn yellowish-green.

Plants that are short of *potassium* do not have enough energy to grow properly. This means they develop poorly formed roots and weak stems. The edges of older plant leaves

appear burnt, because the plant cannot control the amount of water in the leaves properly. This also makes it difficult for the plant to survive **droughts**. Plants with a shortage of potassium are less resistant to pests and diseases.

Plants that do not get enough *phosphorus* are also spindly, with thin, weak stems and stunted or shortened growth. Older leaves turn a dark bluish-green, and plants may fail to produce flowers, seeds and fruit.

The great GM debate

What is GM?

GM stands for 'genetic modification'. Ever since human beings started to grow crops we have been genetically modifying them. We have done this by cross-breeding plants, to make them more suitable for our needs. For instance, farmers sometimes cross one plant with another to get a bigger and better crop, or to make it more resistant to moulds or pests. By using modern gene technology, genetic modification has been speeded up. The genes of a plant in a laboratory can now be directly altered, rather than doing it over months and years by cross-breeding.

GM debate still rages as supermarkets introduce food labelling

Supermarkets were trying to keep ahead of the game in the great GM debate yesterday when two chains, Foodco and Madsbury's, introduced GM labelling on their own-brand foods. "Our customers want to know what they are eating," said a spokesman for Foodco. "Many are very health and environment conscious," she continued, "This new labelling means they can see at a glance if the product contains any GM food ingredients."

Our science editor, Frances Stein, takes a look at both sides of the GM debate.

GM scientists argue that they are merely doing what farmers have been doing for generations – cross-breeding plants to make better crops, but doing it more quickly and efficiently. On the other hand, **environmentalists** are concerned that when GM crops are grown in the wild, the modified genes might 'escape' into the environment. This could result, for example, in wild **species** becoming resistant to pesticide – creating 'superweeds' which cannot be destroyed with standard herbicides.

Another point in favour of GM foods is that they can have positive health benefits, for instance by increasing the storage-life of foods. Crops can also be made more nutritious;

Monarch butterfly

'golden rice', for example, has increased levels of vitamin A, vital for people who rely on rice as their main food. Against this, environmentalists point out that there are health risks. Recent research has found that GM genes can transfer from food to **bacteria** in the human gut which could result in bacteria becoming more resistant to medicines.

However, scientists point to many more advantages of GM crops. They can produce higher **yields**, which could benefit both farmers and poorer nations that need efficient food production.

They also require less fertilizer, which will reduce environmental pollution. They can even be engineered to tolerate extreme environments – such as heat, cold or salty conditions. This would allow food to be grown in more areas to feed the local people. For instance, plants that grow in **drought** conditions will help feed people living in hot dry areas, such as Ethiopia. In spite of this, however, not all Third World countries welcome GM crops. In the middle of a recent **famine** across southern Africa, Zambia refused to take GM food aid on the grounds that it is too dangerous.

Those in favour of GM are quick to point out that it is not just about yield: it can make crops more resistant to pests and diseases, so that fewer pesticides are used. This means less pesticide getting into the environment — and into our bodies via food. On the other hand, GM critics argue that pest-resistant crops actually produce their own insecticide, and that this will also get into the human **food chain**. They also say that insects may become resistant to the insecticide within these crops, so we would again have to use spray pesticides. In addition, the GM insecticide kills not only the target pest insects, it can adversely affect other insects too. Caterpillars

of the rare monarch butterfly suffered serious side-effects when they fed on leaves dusted with pollen from GM maize.

The final argument in favour of genetic modification is that it allows scientists to transfer certain characteristics from one plant to another. It can even give plants characteristics taken from other organisms. Changes are not limited to food crops – plants can also be engineered to produce medicines and raw materials for industry, which could have huge benefits. Yet this same argument is used against genetic modification. It is the ability of genes to move from one species to another that worries GM critics.

There is little doubt that GM has the potential to help us feed the world efficiently, with fewer environmental consequences, if it is developed responsibly. Nevertheless, its critics argue that long-term effects of genetic modification are not sufficiently understood, and that GM organisms need long and careful testing before being released into the environment.

The effects of fertilizers

Although plants receive a natural supply of **nutrients** from organic matter and soil minerals, crop plants usually need extra supplements. Fertilizers are substances containing nutrients needed for plants to grow and remain healthy. Amateur gardeners often add small amounts of fertilizer to the soil, but market gardeners and agricultural farmers use far more. They use them on crops such as wheat, tomatoes and carrots. The most common nutrients added by fertilizers are nitrogen, phosphorus and potassium.

Fertilizers can be divided into two main types:

- **organic** – made from animal or plant products such as manure, fish and bone meal, seaweed and compost

- **chemical** – artificially manufactured to give plants specific amounts of potassium, nitrogen and phosphorus.

There are both advantages and disadvantages in the large-scale use of chemical fertilizers in agriculture.

Fertilizers are used to increase crop **yield**, and are particularly important in large-scale intensive farming.

Advantages

- provide the nutrients plants need to grow, develop and provide good crop yield

- put the nutrients back in soil once crop has been harvested

- do not pollute environment if used properly – improved methods of use reduce waste

- help prevent **famine** in Third World by improving food crop yields – provide food for 2.4 billion people worldwide

- (chemical fertilizers) cost-efficient way of ensuring high crop yield

- (chemical fertilizers) can be made to a special **formula** to provide for specific needs of crop.

Disadvantages

- (chemical fertilizers) do not replace organic matter needed to keep soil structure healthy

- too much fertilizer washed away by rain into rivers and lakes which can cause excessive growth of **algae,** causing pollution and using up oxygen

- harmful nitrates (from nitrogen fertilizers) can leak into drinking water – may cause cancer

- massive amount of energy needed to produce nitrogen fertilizers – using up world's energy reserves

- fuel used to produce fertilizer gives off carbon dioxide (greenhouse gas) into atmosphere

- fields treated with chemical fertilizer give off nitrous oxide (greenhouse gas)

- natural potassium and phosphate reserves will run out in 100 years if people carry on using them as fertilizers.

Fertilizers: bane or boon?

Gulf of Mexico Dead Zone blamed on fertilizer run-off

The US Environmental Protection Agency (EPA) today released a report concluding that fertilizer being washed into the Mississippi River has caused oxygen levels in the Gulf of Mexico to fall too low to support marine life. The Mississippi flows into the Gulf near New Orleans and has created an 8000 square mile* 'dead zone' along the northern coast of the Gulf.

When large amounts of fertilizer **nutrients** are washed off agricultural land and into waterways by the rain they can speed up the growth of microscopic plant-like organisms called **algae**. These periods of rapid growth are called 'algal blooms'. When the organisms die, their decomposition uses up the oxygen in the water needed by other plants and animals. In addition, some **species** of algae produce toxins which poison the water. The area may take years to recover.

The only way to stop this **pollution** is to reduce the amount of chemical fertilizers being used on agricultural land or stop their use completely in areas near waterways. One way of reducing the problem would be to use natural organic fertilizers instead of chemical ones.

In the Gulf, the EPA is leading an initiative, involving nine states and two native tribes along the Mississippi River, to reduce the size of the polluted zone by cutting the amount of nutrients, including nitrogen, entering the Gulf, by 30%. The EPA also recommends that the government should expand programmes that pay farmers to reduce their use of fertilizers.

* US measure: 8000 sq. miles = approx. 20,720 sq. km

Miracle Ferto

The fertilizer you can use all year round

Miracle Ferto will keep your lawn healthy and beautiful from the moment it's laid down. Simply apply **Miracle Ferto** All-Year-Round Fertilizer when you lay your lawn and every month to two months thereafter.

Miracle Ferto All-Year-Round Fertilizer is:

- Formulated specifically for family garden lawns.
- Rich in nitrogen to keep your lawn lush and green.
- Three fertilizers in one:

    1 **Revives** – rapidly greens faded lawns with its high-nitrogen **formula**.

    2 **Maintains** – keep lawns lush and green all year round by providing the right balance of nutrients whatever the time of year.

    3 **Feeds** – even in cooler weather it feeds your lawn, preventing yellowing.

Miracle
Ferto

All year
round
fertilizer

You are what you eat

Our bodies rely on the food we eat to grow, stay healthy and repair themselves. Different animals have adaptations to find and digest different foods.

Zebra

Herbivores

- eat plants which can be difficult to break down and digest

- usually have grinding teeth

- have specialized digestive systems to extract as much nutrition as possible from plant tissues

- tend to eat continuously as plants such as grasses are low in **nutrients**

Killer whale

Carnivores

- feed on other animals

- have cutting and slicing teeth

- swallow food whole or in chunks (unlike herbivores, no digestion takes place in the mouth)

- have large stomachs which produce a lot of acid to digest lumps of meat and bones rapidly (meat is more easily broken down into absorbable nutrients than plants)

- have shorter intestines which quickly absorb nutrients

- tend to gorge food, then fast

Omnivores

- varied diet – can include meat, plants, fungus, fish, fruit and nuts
- front teeth have sharp cutting edges to tear meat
- back teeth grind food to break down vegetable matter
- digestion of plant material begins in the mouth
- have large, single-chambered stomachs
- intestines are longer than carnivores' but not as long as herbivores'
- tend to eat little and often

Brown bear

Scavengers and detrivores

There are also:

- **Scavengers**, such as hyenas and vultures, which are essentially carnivores who rarely kill for themselves but who 'clean up' other carnivores' leftovers.
- **Detritivores**, mainly **invertebrates** such as beetles and earthworms that feed on decaying organic matter on the surface or in the soil.

Vultures

Which of the above categories do humans best fit into?

Chains and pyramids

Energy cannot be destroyed but it can be passed on. That is what happens to the Sun's energy. It is absorbed by plants (producers) during **photosynthesis**, then passed on to the animals (consumers) that eat them, and on again to the animals that eat those animals – and so on. But at each level some of the energy is needed by the organisms and some is lost as heat, so there is less and less energy being passed up the chain. This means that fewer animals are supported at each level. This is sometimes shown in a diagram known as a 'pyramid of numbers'.

The feeding relationships between organisms can also be shown in a **food chain** like these below.

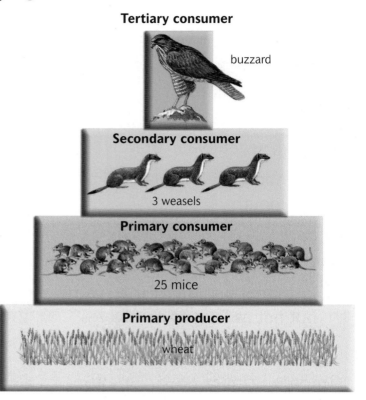

Pyramid of numbers

This is a particularly long food chain.

sunlight   phytoplankton   zooplankton   fish   squid   killer whale

Many have only three links, such as:

sunlight   grass   cattle   humans

What do the arrows in these food chains and webs represent?

Feeding relationships between plants and animals are always more complicated than just a single chain. For instance, in the first chain (opposite), the buzzard could also eat the mouse, and the weasel might eat a beetle which had also fed on the wheat. Other **predators** such as foxes might prey on the mice. More complicated feeding relationships can be shown by drawing a **food web** (right).

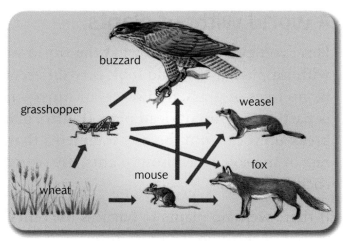

buzzard grasshopper weasel fox mouse wheat

How many food chains can you identify in this food web?

Food chains Down Under

A young **botanist** called Joseph Banks, who sailed around the world with Captain Cook, wanted to study the wildlife of the exciting new lands they visited. The following is an extract from the journal that Joseph made in Australia. Excuse his bad spelling, the dictionary had only just been invented!

A food chain can be constructed from the plants and animals he describes. Remember, humans are part of food chains too.

'Of plants in general the Countrey afforded a far larger variety than its barren appearance seemd to promise. Quadrupeds* we saw but few and were able to catch few of them that we did see. The largest of these was calld by the natives Kangooroo. Besides these Wolves** were seen by many of our people. In the rivers were ducks who flew in large flocks. Land birds were crows, parrots, cockatoes, pidgeons and beautiful doves. Most of these were extremely shy so it was with dificulty that we shot any of them. We saw a white cockatoe in the possession of an Indian†.'

*   'Quadrupeds' are animals with four legs, e.g. cows
** 'Wolves' would have been dingoes.
†   'Indian' was Banks's name for the local aboriginal people

Joseph Banks, the naturalist, who sailed around the world with James Cook in the *Endeavour* (1768–71).

Joseph Banks's 'kangooroo'

The cycle of life

A world without plants

Have you ever wondered what the world would be like without plants? It would look very different. There would be no trees or grass, no flowers or moss – nor any soil (soil is partly made up of dead plants). Also, plant roots help break up rocks into the tiny fragments that make up the rest of the soil, so without plants, the world would be bare rock and sand.

If there were no plants to harness the Sun's energy, there would be nothing for most animals to eat. Since plants provide oxygen when they **photosynthesise**, the atmosphere would be very different too. No food to eat, not enough oxygen to breathe – a world without plants would also be a world without most of the animals we know today.

Barren desert – a world without plants.

Bright flowers and nectar attract pollinating insects.

A world without animals

Now imagine the world with plants, but without animals. A lot of the plants would be very different from the ones we know. Many plants rely on animals to carry pollen between them, so there would be no need for gaudy flowers or attractive scents to encourage insects to visit the flowers and **pollinate** them. There would be no need for seeds with hooks, such as cleavers, that stick to animals' fur. There would be no need for tempting fruit for animals to eat, so that they spread the indigestible seeds in their droppings some distance from the parent plant. Plants would have to rely on other means, such as the wind, to carry their pollen around.

Hooks on burdock seed heads attach to animal fur.

Fern spores are spread by the wind.

Sharp thorns discourage browsing animals.

Without animals, plants would no longer need defence mechanisms to stop them being eaten – there would be no stinging nettles or poisonous berries or sharp thorns. They would also find it more difficult to extract **nutrients** from the soil. Many of the nutrients plants need come from the decayed remains of dead animals, and the soil is kept **aerated** by millions of tiny **invertebrate** animals constantly moving around within it. A world without animals would mean completely different plant life, and probably far less variety.

The everlasting cycle

Our world has evolved (developed) with both plants and animals, and they are adapted to a life of coexistence. Their survival is linked in a cycle of production and decay that recycles vital elements such as carbon and nitrogen.

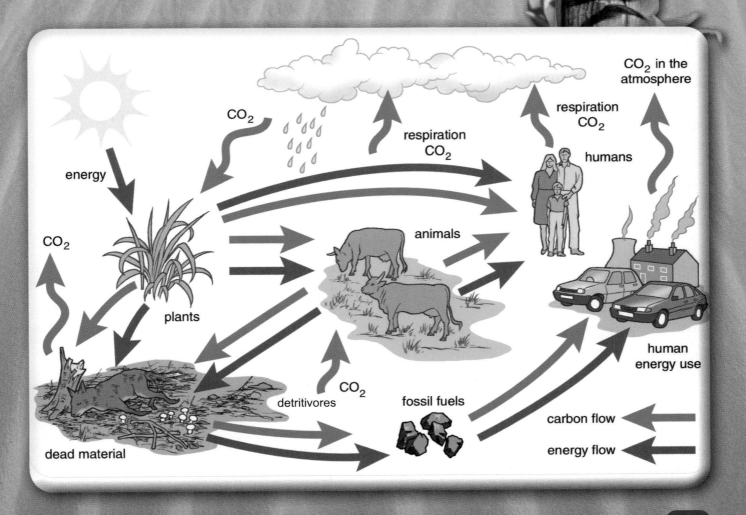

Identification keys

The variety of life on earth is staggering. There are 350,000 known species of beetle alone, and the actual figure may be twice that. In just one square metre of grassland you could expect to find dozens of different plant and **invertebrate species**. With such an array of species we need to be able to tell them all apart. Some species are very similar to each other, so this is not always easy, but scientists have worked out a system of 'identification keys', which unlock the mysteries of sorting out which plant or animal is which.

Take a group of objects. It could be anything; food, clothes or things that you find in your pencil case. Look at the objects carefully and pretend you don't know what they are. What features would you choose to identify them?

- What things about them are the same?
- What things are different?

You could choose:

- what they are made of
- what they are used for
- colour
- texture
- shape

You can now use these features to group your objects together and identify them.

Once you have grouped a set of objects in this way, it is easier to talk about the things they have in common, and the things that are different.

Leaves needlelike —
Leaves not needlelike

Spider keys

The diagram above shows a simple key for the leaves pictured. It uses *shape* as the main way of identifying the leaves. But it could also have used features such as size, colour or texture. Choose one of the leaves and answer the questions about it. Follow the lines to the next question until the key tells you what tree the leaf comes from. This visual key is called a 'spider' or 'tree' key, but it can take up a lot of room, especially if you are identifying a large group of things.

*Leaf-shape glossary

compound	made up of several parts
lobed	rounded like an ear
palmate	hand-shaped
pinnate	having leaflets ranged along each side of the stem
serrated	notched edge

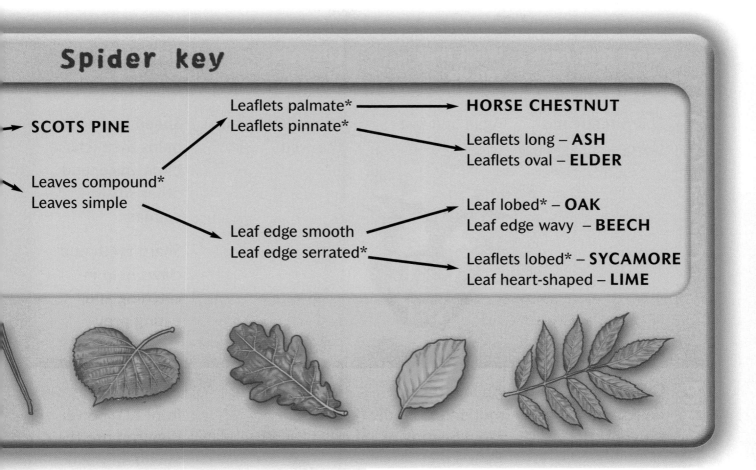

Spider key

SCOTS PINE

Leaves compound*
Leaves simple

Leaflets palmate* ──────→ **HORSE CHESTNUT**
Leaflets pinnate*

Leaflets long – **ASH**
Leaflets oval – **ELDER**

Leaf edge smooth
Leaf edge serrated*

Leaf lobed* – **OAK**
Leaf edge wavy – **BEECH**

Leaflets lobed* – **SYCAMORE**
Leaf heart-shaped – **LIME**

Numbered keys

To save space, a spider key can be turned into a numbered key. The numbered key for the the spider key above would look like the diagram on the right.

Constructing a numbered key

A key contains a series of questions designed to identify an object that you cannot name. The questions are in pairs and you choose the right answer to lead you on to the next pair of questions. At the end of the questions, you end up with the name of the animal or plant you are trying to identify. This is called 'keying out'.

Numbered leaf key

1a leaves needlelike	**Scots pine**
1b not needlelike	go to 2
2a leaves simple	go to 3
2b leaves compound*	go to 4
3a leaf edge smooth	go to 5
3b leaf edge serrated*	go to 6
4a leaflets palmate*	**Horse chestnut**
4b leaflets pinnate*	go to 7
5a leaf lobed*	**Oak**
5b leaf edge wavy	**Beech**
6a leaflets lobed*	**Sycamore**
6b leaf heart-shaped	**Lime**
7a leaflets long and thin	**Ash**
7b leaflets oval	**Elder**

Adapt or die

	What they have	What they can do	Lifestyle	Behaviour
Animal: Caracal (cat)	Sharp teeth and claws	Agile and fast	Hunters when in the wild	Speed and agility, helps them chase birds, mice, and other small animals Sharp teeth and claws help in catching and eating **prey**
Animal: Hummingbird	Long narrow beaks and tongues	Ability to hover motionless by the rapid beating of their wings	Feed on flower nectar	By hovering in front of a flower they can use their long beaks and tongues to probe deep into flowers to extract the nectar
Animal: Arctic hare	Fur turns white in winter Broad feet help them run on soft snow	Agile and fast	Lives where there is always heavy snow during winter months	White coat provides **camouflage** so it is not so obvious to **predators** (e.g. wolves and lynxes)

These animals show adaptations to their environment and lifestyle. Charles Darwin and others had noticed adaptations such as these and were curious to know how they had occurred. Here are two 19th-century schools of thought, one championed by a French scientist called Jean-Baptiste Lamarck the other by Charles Darwin, Alfred Russel Wallace and others. The pictures below illustrate how their arguments went. Who do you think was right?

A Lamarckian explanation

B Darwinian explanation

Explanation B is correct. The giraffe with the feature that helped it survive, its long neck, is more likely to live long enough to breed and pass on that feature. Explanation A is incorrect. The giraffe's baby would not inherit the mother's stretched neck. We do not pass on features that we have acquired during our lifetime, only things that we are born with. Of course, in reality changes take place over many generations. These cartoons only show you the basic idea.

Evolution

Times of change

Charles Darwin was a scientist who lived in the nineteenth century. At the time he was living and working there was still much scientific and religious debate about how life on Earth has changed over time. Fossils proved that some animals that lived in the past no longer exist today. But did animals and plants change – evolve – into new forms or did new **species** simply appear? Darwin was one of the greatest thinkers on this subject and his ideas formed the basis of what we understand about evolution today.

Natural selection

Darwin realized that there is 'natural variation' among animals and plants of the same species. Just look around you, some of us have brown eyes, others blue or grey or green. Some of us are tall others short, and so on. He reasoned that if one of these natural variations also happens to give the individual an advantage and helps it survive long enough to breed, then that useful feature is likely to be passed on to its offspring. With time, such changes might accumulate and lead to the formation of a new species. This process is called 'evolution by natural selection'. Darwin explained it in his famous book on the subject, *On the Origin of Species by Natural Selection,* published in 1859.

Revolutionary ideas

While many scientists were excited and accepting of Darwin's theory, some of his ideas shocked the general public. Religion was a very important part of Victorian society and some felt his idea that evolution was random and that human beings and apes shared a common ancestor challenged their religious beliefs. A few people still took the Bible literally, and were outraged by his ideas. However, the growth of sciences such as biology and geology meant that Darwin's basic ideas were soon accepted by the scientific and educated community.

Natural selection has acted in nature, in modifying and adapting the various forms of life to their several conditions and stations.

Charles Darwin

Two extracts from *On the Origin of Species* by Charles Darwin

In order to make it clear how natural selection acts let us take the case of a wolf which preys on various animals and let us suppose that the fleetest [*fastest*] **prey**, a deer had increased in numbers and that other, [*slower*] prey had decreased in numbers, during that season of the year when the wolf is hardest pressed for food. The swiftest wolves would have the best chance of surviving. Some of its young would probably inherit the same habits or structure, and by the repetition of this process, a new variety might be formed which would supplant [*replace*] the parent-form of wolf.

More individuals are born than can possibly survive. A grain in the balance will determine which individual shall live and which shall die, which variety or species shall increase in number, and which shall decrease, or finally become **extinct** ... The slightest advantage in one ... over those with which it comes into competition, or better adaptation ... to the surrounding physical conditions, will turn the balance.

The dodo became extinct after pigs were introduced to the island of Mauritius, because they destroyed its eggs and competed with it for food.

Although these two quotations show that he understood the basic principles behind evolution and adaptation, Darwin did not know how it worked. It was not until well into the 20th century that scientists began to properly understand how **genes** (chemical codes that make up our **chromosomes**) determine what we are like.

The life of Charles Darwi

Darwin made extensive collections and observations on his voyage on HMS *Beagle*.

February 12
Charles Robert Darwin born in Shrewsbury, England.

December:
Went to Cambridge to study for the clergy – spent free time collecting beetles and socializing.

September 7:
After visiting several places in South America HMS *Beagle* set sail for Galapagos Islands.

1809

1827

1835

1825

1831

1837

Went to University of Edinburgh to study medicine. Hated sight of blood – didn't do well.

Invited to be naturalist aboard HMS *Beagle* on two-year survey of South America.

Returned to London – began to develop theory of evolution using evidence he collected on voyage.

HMS *Beagle*

Darwin saw that each Galapagos finch **species** filled a specific island 'niche'.

During the debate, there was much commotion in the audience (one woman fainted!), with both sides claiming to have won the argument.

January 24:
Married cousin, Emma Wedgwood. (They had 10 children).

May:
The Voyage of the Beagle published.

June 30:
Heated debate about Darwin's theory at Oxford.

Published *Descent of Man* about human evolution.

1839

1860

1871

1859

1867

1882

November:
Finally published his theory in *On the Origin of Species by means of Natural Selection.*

Despite initial opposition from the Church, Darwin's theory of evolution accepted in scientific circles throughout Europe.

Died. Buried at Westminster Abbey.

Darwin's book virtually sold out on the first day.

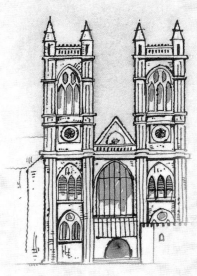

Finding your niche

Living things are found all over the world, from Arctic wastes to hot deserts, from lush rainforests to dark caves, in the air, the water and in the soil. These different places are called **habitats**, and the conditions that exist within habitats, such as temperature, altitude and humidity make up the environment. Animals and plants show adaptations to their environment and each occupies a specific role in that environment with relation to the other animals and plants that live there. This role is called an 'ecological niche'. It includes:

- how the animal or plant uses the resources in the environment to survive

- how it affects, and is affected by, other living things in the environment.

Most **species** on Earth occupy different ecological niches within their environments. Where two animals occupy the same niche, there will be problems. Eventually the species best adapted to its position in the community will probably drive out the other.

Siskin

Same habitat, different niche

Crossbills and siskins

Similar animals can live in the same habitat but occupy a different niche because they feed on different things.

The crossbill lives in pine forests. Its unusual crossed-over beak is an adaptation to feeding on pine cones. The scissor action of its bill allows it to cut open the cones and get at the seeds.

Crossbill

The siskin is related to the crossbill (they are both finches) and it also lives in pine forests. However, although it is also a seed eater, it feeds on other tree seeds, as well as conifer seeds, and also eats buds, meaning that it is not in direct competition with the crossbill for food and therefore occupies a different ecological niche.

Same niche, different continent
Thompsons gazelles and kangaroos

The Thompsons gazelle is a grass feeder. It lives in the African savannah and has a long face and jaw which house the rows of grinding teeth it needs to chew grass. It is fast, so that it can run from **predators**, such as lions and leopards.

Occupying the same niche in another part of the world, Australia, is the kangaroo. It also lives in open grasslands and feeds on grass. It shows similar adaptations to its environment and lifestyle as the gazelle. It, too, has a long face and jaw with grinding molar teeth and moves swiftly to escape predators such as the dingo. Unlike the Thompsons gazelle, however, it doesn't run, it hops, at speeds up to 48 km per hour.

Same niche, better adaptation
Grey squirrels and red squirrels

Grey squirrels, introduced into Britain from America in 1876, overlapped the same ecological niche as the native red squirrel. They ate the same food, nested in the same trees and were eaten by the same predators. Unfortunately for the red squirrel, the grey squirrel proved to be a much more adaptable feeder and can thrive on acorns and a wide variety of other foods, whereas the red squirrel prefers pine cones. In many areas the more **versatile** grey squirrel replaced the red squirrel, particularly in mixed woodlands, and today the red squirrel is restricted to pine forests in the Scottish Highlands and other more remote areas.

Thompsons gazelle

Kangaroo

Red squirrel

Grey squirrel

Endangered species

Lifestyle changes

Many things can happen to change the environment in which animals and plants live. Some changes are natural, such as some variations in climate; others are caused by humans, such as deforestation or the introduction of new **species**. When the environment changes, some species lose out, others adapt – sometimes in surprising ways.

A young polar bear scavenges among rubbish left by humans in its Arctic environment.

The fate of a vegetarian carnivore

The giant panda has filled a niche by feeding almost entirely on bamboo shoots and roots, a food no other animal eats. However, the panda was originally a carnivore. Its teeth and digestion are adapted to eating meat. This means that bamboo is a very inefficient food for it. The panda needs to consume up to 18 kg of bamboo roots and shoots a day in order for its digestive system to extract enough energy and **nutrients** for it to survive. To do this, pandas feed for 10–12 hours every day. The panda is also at risk from environmental change. Bamboo forests are under threat of deforestation and bamboo plants also have a natural lifecycle of about 100 years, when all plants in an area die back almost completely, and the panda is left with virtually no food. Because the panda cannot cope with these changes, for instance by finding a new source of food, it may become **extinct**.

One of the few giant pandas living in the wild feasts on bamboo.

The Red Dragon Hotel
Beijing
China

23 March 2003

Dear Callum

Wow, it's fantastic here in China. Yesterday we walked along the Great Wall and Dad says we are going to see the terracotta soldiers next week. But the thing I really wanted to do was visit the Wolong Nature Reserve where you can see giant pandas in the wild. But here's some stuff I've found out about pandas and now I'm not so sure we should go. What do you think?

Hey, did you know there are only 1000 giant pandas left in the wild? They eat mainly bamboo but the bamboo forests where they live are being destroyed to make room for people to live. There are over one billion people in China and they need lots of room too. The Chinese government has created 11 nature reserves in bamboo forests but pandas are still in danger. They breed very slowly so they can't increase their numbers quickly and even in reserves they are poached and killed for their fur. A single pelt can fetch US$200,000 — that's nearly £300,000! Poachers get life sentences if they're caught, but for that money, people still risk it.

The government created the Wolong Nature Reserve in 1975. It was supposed to be like a panda heaven where they would be safe. Loads of money was put into the reserve by conservation organizations such as the Worldwide Fund for Nature, but now Wolong has become a victim of its own success. Towns and villages have sprung up in the reserve itself and the local human population has increased by 70%. Each tourist who comes to see the pandas causes their habitat to shrink a little more. Tourists stay in hotels, eat local products and buy local souvenirs, which all uses up land and resources. I didn't think we would have an effect on the pandas by visiting Wolong, but it seems tourists damage the local environment too. It's hard to think that we might harm the pandas if we go to see them. Lots of zoos are trying to breed pandas in captivity, so perhaps we'll just go and see them at Beijing zoo instead.

See you next term.

Bye for now

Tom

Adapting to your

The Arctic tundra

Tundra is the name for cold, treeless areas where the soil remains frozen all year long. This is called 'permafrost' and can be up to 700 m deep. Tundra covers one fifth of the Earth's surface and can be found in the far north of Canada, Russia, Scandinavia and Alaska, as well as mountain regions. Plant roots cannot penetrate the tundra permafrost which makes it impossible for tall trees to grow here. The animals and plants that live in the tundra are adapted to life in a cold climate in many ways.

One of the main concerns of animals is keeping warm and they often display one or more of these adaptations:

- short, stocky limbs and small ears with a relatively small surface area for heat to be lost from

- thick feathers or fur for insulation, keeping in warmth and sealing out the wind and cold

- a thick fat layer gained quickly during spring and summer that provides insulation and fuel during the winter

- adaptations to prevent bodily fluids freezing (insects have a type of antifreeze in their blood)

Camouflage is important for both hunters and hunted: many animals and birds change the colour of their feathers or fur from brown or mottled in the summer to white in the winter.

Arctic hare

Snowy owl

Wolf

Caribou

Woodchuck

Lynx

Arctic fox

environment

Snowy owl

Caribou

Wolf

Musk ox

Lynx

Arctic hare

Woodchuck

Living in Madagascar

The Madagascan rainforest

Madagascar is a large island off the eastern coast of Africa. It is home to many unique animals and plants, not found anywhere else in the world, that are adapted to life in the island's various **habitats**. It has many different environments ranging from desert to rainforest. Conservation is an important issue in Madagascar and many of its **species** are endangered.

Aye-aye – this lemur feeds on fruit and on grubs which it fishes out of rotting wood with its specially thin and flexible middle finger.

Ring-tailed mongoose – feeds on small mammals and reptiles, including snakes.

The hedgehog-like tenrec – eats insects, worms, fruit and small amphibians.

Giant hissing cockroach – feeds on detritus on the forest floor.

Souimanga sunbird – feeds on nectar.

Fossa – Madagascar's largest carnivore looks like a cat, but is in fact related to the mongoose.

Ring-tailed lemur – feeds on fruit and occasionally leaves in the Madagascan forest.